INCOBAT: Innovative Cost Efficient Management System for Next Generation High Voltage Batteries

RIVER PUBLISHERS SERIES IN TRANSPORT TECHNOLOGY

Series Editors

HAIM ABRAMOVICH
Technion – Israel Institute of Technology
Israel

THILO BEIN
Fraunhofer LBF
Germany

Indexing: All books published in this series are submitted to Thomson Reuters Book Citation Index (BkCI), CrossRef and to Google Scholar.

The "River Publishers Series in Transport Technology" is a series of comprehensive academic and professional books which focus on theory and applications in the various disciplines within Transport Technology, namely Automotive and Aerospace. The series will serve as a multi-disciplinary resource linking Transport Technology with society. The book series fulfils the rapidly growing worldwide interest in these areas.

Books published in the series include research monographs, edited volumes, handbooks and textbooks. The books provide professionals, researchers, educators, and advanced students in the field with an invaluable insight into the latest research and developments.

Topics covered in the series include, but are by no means restricted to the following:

- Automotive
- Aerodynamics
- Aerospace Engineering
- Aeronautics
- Multifunctional Materials
- Structural Mechanics

For a list of other books in this series, visit www.riverpublishers.com

INCOBAT: Innovative Cost Efficient Management System for Next Generation High Voltage Batteries

Editors

Eric Armengaud

AVL List GmbH
Austria

Riccardo Groppo

Ideas&Motion S.r.l.
Italy

Sven Rzepka

Fraunhofer Gesellschaft zur Förderung der angewandten Forschung E.V.
Germany

River Publishers

Published, sold and distributed by:
River Publishers
Alsbjergvej 10
9260 Gistrup
Denmark

River Publishers
Lange Geer 44
2611 PW Delft
The Netherlands

Tel.: +45369953197
www.riverpublishers.com

ISBN: 978-87-93519-63-3 (Hardback)
978-87-93519-62-6 (Ebook)

Contents

List of Figures

List of Table

List of Abbreviations

ADC   Analog to Digital Converter
ASW   Application Software
BCI   Bulk Current Injection
BSW   Basis Software
CAN   Controller Area Network
CCU   Central Control Unit
CPU   Central Processing Unit
DAC   Digital to Analog Converter
DPI   Direct Power Injection
EIS   Electrochemical Impedance Spectroscopy
EMC   Electromagnetic compatibility
FE   Finite Element
FEV   Fully Electric Vehicle
FFT   Fast Fourier Transform
FMEA   Failure Mode and Effects Analysis
HV   High Voltage
HW   Hardware
PCB   Printed Circuit Board
SoC   State of Charge
SoF   State of Function
SoH   State of Health
SW   Software
TI   Technical Innovation

Executive Summary

Electro-mobility is considered as a key technology to achieve green mobility and fulfil tomorrow's emission standards. However, different challenges still need to be faced to achieve comparable performances to conventional vehicles and finally obtain market acceptance. Two of these challenges are vehicle range and production costs. In that context, the aim of INCOBAT (October 2013–December 2016) was to provide innovative and cost efficient battery management systems for next generation HV-batteries. INCOBAT proposes a platform concept that achieves cost reduction, reduced complexity, increased reliability and flexibility while at the same time reaching higher energy efficiency.

- Very tight control of the cell function leading to a significant increase of the driving range of the FEV.
- Radical cost reduction of the battery management system with respect to current solutions.
- Development of modular concepts for system architecture and partitioning, safety, security, reliability as well as verification and validation, thus enabling efficient integration into different vehicle platforms.

The INCOBAT project focused on the following twelve technical innovations (TI) grouped into four innovation groups (see Figure 1):

1

Customer needs and integration aspects: These innovations ensure a correct identification of customer needs and enable efficient integration into different platforms.

Transversal innovation: This second group targets the optimisation of the system architecture and its consistent definition in the technologies and in the system hierarchies. The focus was set on providing a consolidated basis to simplify later industrialization of the proposed technologies.

Technology innovation: This third group aims at improving the components of the E/E control system, including topics such as smart sensors, innovative computing platforms or control strategies.

Transversal innovation: This last group targets the evidences related to the trust in the technical solutions with respect to correct operation, functional safety, security and reliability. This group of technical innovations is an indicator for the maturity of the proposed technology and further provides information on the efforts required for proper integration and validation of the system.

The main INCOBAT technical achievements can be summarized as follows:

- Improving the range of the electric vehicle by better use of the electrical energy stored within the battery, realized by a combination of TI01 (mission profiles), TI03 (efficient partitioning), TI05 (multicore computing platform), TI06 (smart module management unit) and TI08 (improved BMS control algorithms).
- Significant decrease of costs for BMS hardware, realized by a combination of TI03 (efficient partitioning), TI04

(integration of multiple functionalities within the same control unit), TI05 (multicore computing platform), TI06 (smart module management unit), TI07 (modular SW platform) and TI08 (improved BMS control algorithms).

- Provide modular concepts for efficient integration into the vehicle, realized by TI02 (model-based systems engineering), TI03 (efficient partitioning), TI05 (multicore computing platform), TI07 (modular SW platform), TI09 (safety and security co-engineering), TI10 (design and validation plan) and TI11 (reliability and robustness validation).

Achievements regarding dissemination and exploitation of the INCOBAT outcomes include 21 peer-reviewed publications and a dedicated cluster workshop to exchange information between related projects, as well as the development of a dedicated exploitation plan and sustainability model should be highlighted.

INnovative COst efficient management system for next generation high voltage BATteries (INCOBAT)

1 Introduction

In recent years, electric mobility has been promoted as the clean and cost-efficient alternative to combustion engines. Although there are solutions available in the market, mass take-up did not yet happen. There are different challenges that impede this process from an end user's point of view such as cost of the vehicle, driving range, or infrastructure support. Several of these challenges are directly connected to the battery, the central element of the full electric vehicle (FEV). The cost of the battery accounts for up to 40% of the total cost of an FEV, while the range of the FEV is fairly limited in comparison to the combustion engine.

The aim of the INCOBAT project (October 2013–December 2016) was to provide an innovative and cost-efficient battery management systems for next generation of HV-batteries. INCOBAT proposes a platform concept that achieves cost reduction, reduced complexity, increased reliability, and flexibility while at the same time reaching higher energy efficiency.

The consortium consists of the partners AVL List GmbH (coordinator), Ideas&Motion, Fraunhofer ENAS, Infineon Technologies AG and Infineon Technologies Austria AG, Impact Clean Power Technology S.A., Manz Italy Srl and Chemnitzer Werkstoffmechanik GmbH and is in the position to provide a 100% European value chain for the development of next generation HV battery management systems.

The Main Objectives of the INCOBAT Project

- Very tight control of the cell function leading to a significant increase of the driving range of the FEV.
- Radical cost reduction of the battery management system compared to current solutions.
- Development of modular concepts for system architecture and partitioning, safety, security, reliability as well as verification and validation, enabling efficient integration into different vehicle platforms.

The INCOBAT project focused on the following 12 technical innovations (TI) grouped into 4 innovation groups (Figure 1):

Customer needs and integration aspects: These innovations ensure a correct identification of customer needs and enable efficient integration into different platforms. This was supported by the use of mission profiles (TI-01) – in order to take the different driving styles of the customers, the different traffic conditions in the same scenarios and the different tracks into account by the integration into a demonstrator vehicle (TI-12).

Transversal innovation: This second group targets the optimisation of the system architecture and its consistent definition in the technologies and in the system hierarchies. The focus was set on providing a consolidated basis in order to simplify later industrialization of the proposed technologies. This includes TI-02 "Model-based systems engineering" to improve correctness, completeness and consistency of system specifications, TI-03 "System architecture – efficient partitioning of the functionalities" for system optimization at BMS or even vehicle level and TI-04 "Integration of multiple functionalities" to reduce the number of electronic control units (and thus related costs) in the vehicle.

Technology innovation: This third group aims at improving the components of the E/E control system. Regarding the electronic parts, it includes TI-05 "TriCore AURIX^{TM} Platform for additional computing resources" and TI-06 "Smart and integrated module management unit". On the software side, this was achieved by TI-07 "Modular SW platform" and by TI-08" Improved BMS control algorithms".

Transversal innovation: This last group targets the evidences related to the trust in the technical solutions with respect to correct operation (TI-10 "Design and validation plan including reliability consideration"), functional safety and security (TI-09 "Definition and integration of safety and security concept") and reliability (TI-11 "Reliability and robustness validation"). This group of TIs is an indicator for the maturity of the proposed technology and further provides information on the efforts required for proper integration and validation of the system.

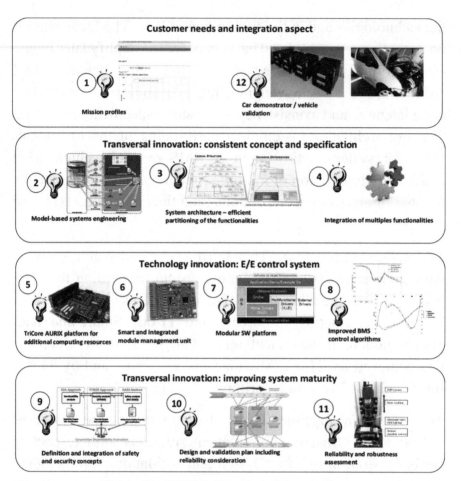

Figure 1 INCOBAT technical innovations.

From the technical point of view, INCOBAT consisted of three innovation work packages (WP1, WP2, and WP3), one implementation (WP4) and one evaluation (WP5) work package (Figure 2). The purpose of the innovation WPs was to provide innovative outcomes, either as transversal innovation over the different technologies (regrouping requirements, system partitioning, safety, security, test planning), or as technology innovation (control strategy, SW, and HW). In the implementation,

WP and the evaluation WP, the technology was integrated as demonstrator and evaluated according to performance, safety/ security or reliability aspects. WP6 focused on project management and sustainability of project results.

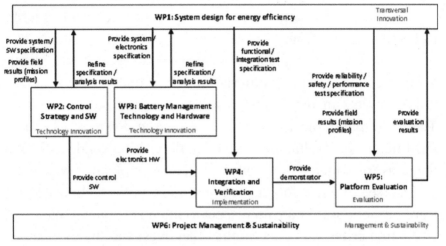

Figure 2 INCOBAT project structure.

The remaining topics are discussed in the following sections.

2 System Design for Energy Efficiency (WP1)

The objective of this work-package was the definition of the next generation modular BMS platform, integrating different topics that have to be considered across the entire supply chain. It aimed at providing consistent and structured information in order to combine different disciplines more tightly and ensure consistency of the information exchange. This included the implication of all underlying technologies (like CPU, system and software architectures) and methodologies (like system modelling, ISO 26262) for requirements engineering, system architecture and partitioning, safety, and security analysis, as well as, cost and reliability engineering.

2.1 Application Scenarios and Requirements

Targets of this activity were the refinement of the project objectives and the systematic identification and management of BMS technical requirements.

The first target (refinement of project objectives and identification of measurement criteria) aimed at making the intended improvements in INCOBAT easier to monitor and to evaluate. The three main objectives have been broken down into 12 technical items. This refinement aimed at describing single technical domains for which an improvement has been planned within the project (see Figure 3 for an example). Each of the technical items were further described with respect to intended content, baseline, evaluation criteria, impact on the objectives, and finally mapped to relevant deliverable.

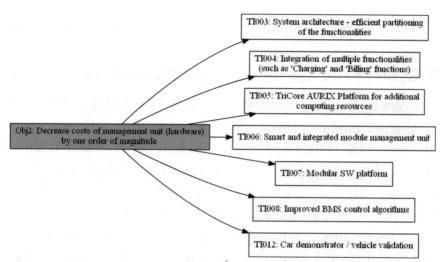

Figure 3 Refinement of project's objectives into technical items: for objective 2 (Decrease costs of management unit).

The second target (systematic identification and management of BMS technical requirements) is linked to the systematic technical description of the BMS. For that purpose, methods (including template) for requirement elicitation and specification have been defined. Parallel to the requirement specification, a system model has been set up. Furthermore, different enhancements for the existing requirement management tool have been proposed to support requirement management during the project.

The methods related to requirement elicitation and specification relies on structuring the requirements according to their abstraction level and their domain. For INCOBAT, the following levels have been defined (Figure 4):

- L1 Powertrain level,
- L2 Battery level,
- L3 BMS central control system, L3 BMS satellite control systems
- L4 BMS central control SW, L4 BMS central control HW, L4 BMS satellite SW, BMS satellite hardware.

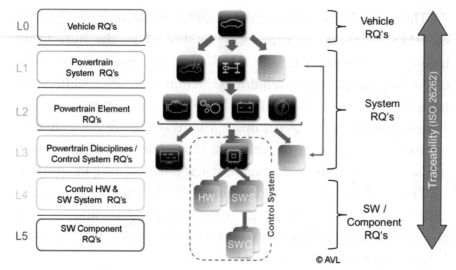

Figure 4   Overview of the abstraction levels.

2.2 System Architecture and Partitioning

This task focused on the systematic description of the BMS control system architecture; further architecture exploration, evaluation, and optimization, focusing on the set-up of the architecture description for the BMS control system according to the requirements were identified previously using the description language SysML.

The description approach follows the concept of the abstraction levels described in task T1.1 (L1–L4 in this project). It introduces the following views in order to provide a more complete description of the system and of its dependencies (Figure 4):

- Logical view: describes the logical structure of the system and its decomposition. This provides an overview of the system and the hierarchy of the sub-components within the system.

- Technical view: describes the technical dependencies in the system. This view focuses on the interfaces between the components.
- Requirement view: depicts the requirements (automatically imported from the requirement management tool) and their mutual dependencies and relations to system components.

These different views enable the description of different aspects of the system and further enable a mapping (establish logical links) between these different aspects. These traces make the relationships between the different aspects explicit.

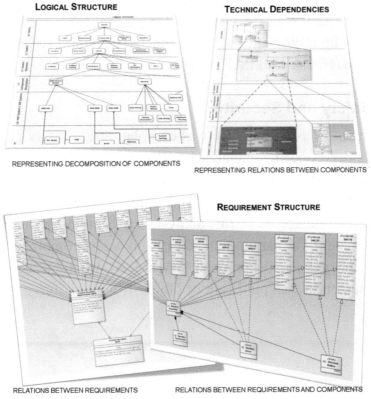

Figure 5　Different views for the modeling of the architecture.

The system architecture specification model used to model the BMS system as well as its direct environment (HV battery, powertrain and vehicle elements as appropriate) resulted in 1132 model artifacts in total and 2206 connections between these elements, including and mapping functional, architecture and safety information.

An important point at this stage was the correct mapping with WP2 (software) and WP3 (electronic hardware) from the vocabulary, expertise and tool point of view. A key aspect was the tool integration, a mapping of system development artifacts with the SW development framework. During the project, different tool interfaces for generation of AUTOSAR aligned SW information where created, e.g., for the operating system (OIL file), for the BSW configuration and for the automated generation of HW–SW interface (HSI acc. to ISO 26262).

2.3 Safety and Security

The main focus of this task was completing the safety analysis and safety concept—enhanced with security features—and to iterate with WP2 and WP3 to consolidate the mapping between resulting safety requirements and implemented safety mechanisms. The outcomes of this analysis has been integrated in the Development Validation Plan (DVP) and a new combined safety and security analysis approach (SAHARA) war subject to some flagship conference publications (Table 1, Publications 5, 9, 10).

The second main outcome of this task was the analysis of the ISO 26262 validation process and optimization of the validation activities with respect to available test environments. Different test environments are available (e.g., Model-in-the-loop for validation of the main SW functions, Software-in-the-loop for validation of the integrated software, Processor-in-the-loop for validation of the firmware on a target processor, Hardware-in-the-loop for validation of the E/E control system, dedicated test-beds for validation of HV batteries or powertrains, and vehicle tests). Each of these environments present a trade-off between realism of the test vectors and system reaction versus degree of controllability and observability. A higher degree of controllability is required to move the system into the desired (possibly faulty) status and a higher degree of observability is required to observe and analyse the system reaction. On the other side, a more mature system (e.g., assembled battery, vehicle) provides a more comprehensive system and respective environment and, therefore, higher confidence on the appropriateness of the test campaign. During this activity, a mapping between the ISO 26262 validation tasks and respective test environment was developed in order to identify the cheapest yet mature enough environments to execute the respective test campaigns and finally increase validation efficiency.

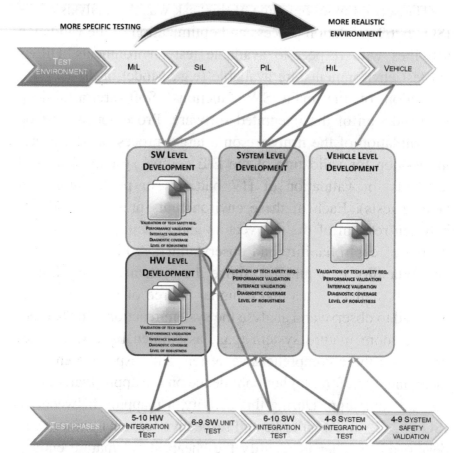

Figure 6   Depiction of ISO 26262 aligned development phases and mapping of respective tests and test environments.

A detailed mapping of safety related test campaigns, ISO 26262 demanded activities and tightly integrated more efficient test environments is depicted in Figure 6. This strengthens the trust in the developed system or enables reaching a high maturity of implemented safety features and ensure a clear vision of further effort required for start of production (SOP) ready product.

2.4 Design and Validation Plan

An important aspect during the development phase was to ensure the alignment between the specification of targets, intended behavior and architecture of the system (WP1), software and hardware implementation (WP2 and WP3, respectively) and systematic validation of the system (WP5). It is clear that the BMS proposed within INCOBAT was relying on different SW and HW components, each providing different maturities. They were ranging from certified chips and industrial control strategies to prototype PCBs and respective SW drivers. Consequently, some components had already been validated outside of INCOBAT, while other components were developed until TRL3-4, therefore, not mature enough for specific test campaigns such as reliability testing.

Consequently, an important target was to provide a comprehensive map of confidentiality level of the INCOBAT solution – what was the maturity before INCOBAT, which maturity had been reached after INCOBAT, and which validation activities will still be required to achieved a given maturity level (e.g., efforts required to move a given INCOBAT technology to SOP).

For this reason, a DVP for the BMS architecture was created. In order to ensure a high consistency with the previous and remaining development work, the DVP was aligned with the system architecture description, all listed measures are directly linked to the logical view levels L0 (vehicle level) to L5 (BMS HW/SW level), as depicted in Figure 7.

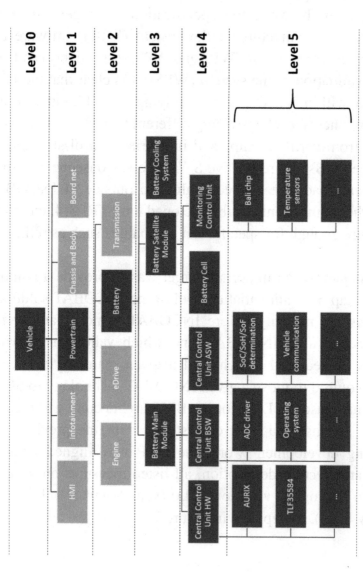

Figure 7 Development validation plan level structure being adapted to the BMS system architecture.

The DVP comprises the following parts: The logical view level with its respective system under test (e.g., L0: vehicle), the type of validation and testing activity, information on the relevance of the respective validation measures w. r. t. the INCOBAT project and a detailed description of the individual validation methods. Figure 8 highlights the mapping of component abstraction levels and the different test aspects, which provide a list of resulting test methods required to execute for a specific system integration level.

Test aspects, which ensure the proper integration of the INCOBAT HV system, were identified in terms of the following test types:

- Functional testing (lab environment): Testing of a product or system with respect to its functional requirements (black-box approach).
- Load/stress testing (realistic environment at normal and anticipated peak load conditions): The goal of a load test was to check whether the system was able to handle normal load conditions or not. In case of stress testing the load exceed the normal usage pattern to investigate the system's response at unlikely load scenarios.
- Fault injection testing (outside of specified environment): Investigation of the (HW or SW) system's response for defined induced faults with the goal of measuring its fault tolerance and robustness.
- Reliability testing (lab and realistic environment at anticipated peak load conditions and above in case of accelerated tests): Testing of the functionality as function of time to determine operating life and or failure modes.

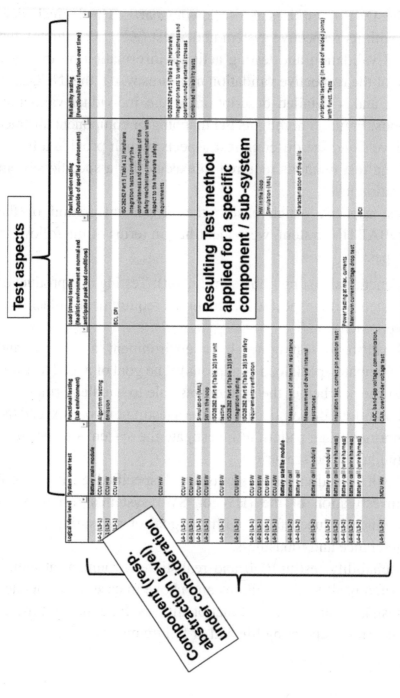

Figure 8 Depiction of DVP approach.

3 Control Strategy and Software (WP2)

Two main targets of the INCOBAT project regarding the software development scope were:

- the delivery of innovative control algorithms to efficiently manage the complexity of the battery;
- the definition, implementation, and integration of a comprehensive software architecture to support the development and to exploit the BMS hardware platform.

Concerning the BMS control algorithms, different challenges were arising during the development and respective validation:

- Complexity of the physical system (battery) to control, and the respective accurate sensing of the environment;
- Scalability to the different chemistries to support;
- Complexity of the entire firmware and need to achieve front-loading (faster iterations even if the entire software is not available).

An important approach of the INCOBAT project was to develop more accurate algorithms to increase battery life but also, more importantly, allow for a predictable battery life. This is essential for creating a viable market for battery leasing and pricing, leading to more business opportunities.

3.1 Definition and Simulation of the State of Function (SoH and SoC) and of the Infrastructure Interfaces

Classical State of Charge (SoC) and State of Health (SoH) estimations algorithms are usually running on module-level due to the high-computational demands. One important target of the research in the project was identifying possible improvements with respect to SoX estimation accuracy, exploiting the additional computing power of the AURIX^{TM} multicore processor. These results have been confirmed through accurate simulations (Figure 9), demonstrating a significant difference between the estimation on cell-level (red curve) and the estimation on module-level (blue curve), especially at the end of the cycle (6000 s).

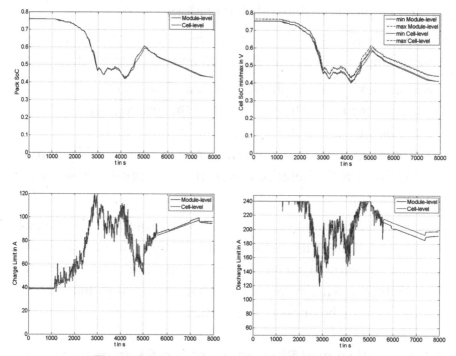

Figure 9   Estimation results for SoC and SoF.

A parallel development was carried out, to investigate the possible use of the Electrochemical Impedance Spectroscopy (EIS) in an embedded environment (Figure 10). Many efforts have been devoted to the development and the validation of an EIS algorithm easily embeddable into the selected microcontroller architecture.

With a better estimation of the battery SoH, the EIS algorithm is expected to deliver:

- reduction of number/size of cells by better battery health management to achieve same lifetime and same mileage;
- added value by advanced diagnosis and predictive maintenance;
- large scale creation of a knowledge database on battery behaviour for further optimization of cell technology;
- 2nd life: higher accuracy of battery, module and cell health assessment, therefore, higher (more accurate) remaining value.

Classic approaches to the EIS problem involve FFT computations, with all the related drawbacks ("leakage" effects at the beginning/end of the measurement window, large amount of data coming from FFT, ...). The research conducted in the INCOBAT project led to an innovative approach, based on the heterodyne method; the solution allows downscaling of the EIS algorithm, enabling the execution on embedded processors, with simple calculations that can be performed on real time.

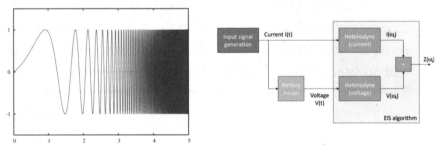

Figure 10   EIS solutions analyzed in the project.

**Enhancement of Active Balancing Functionalities
to Support EIS**

Although most of the research on the EIS has been conducted
with the target of the lab testing, a parallel research branch
was carried out, to investigate possible industrial implementation of the needed HW interfaces inside the chip managing the
active balancing. This analysis revealed that the reuse of some
HW devices, already present in the BALI, with some required
adaptions, could lead to interesting opportunities in serving the
requirements from the EIS algorithm.

3.2 BMS Simulation on Different Configuration and Scenarios

The validation of BMS strategies by means of simulation is an
important aspect for frontloading – the capability to confront
the algorithms with a comprehensive and realistic environment
during an early development stage. This enables a more comprehensive space exploration as well as a faster evaluation of
the proposed algorithms, leading to the development of more
optimized solutions. Therefore, the work focused on three main
topics:

1. Evaluation and validation of the EIS innovative approach
   in a protected, yet realistic environment (Figure 11). This
   assessment has been performed in the test environment, and
   provides measurement analysis of the respective stimulus
   and accuracy of the response, from different cells.

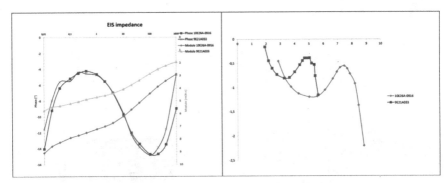

Figure 11 Comparison of EIS impedance measures of different cells.

2. A dedicated simulation environment was provided to simulate and therefore abstract the basic SW and related underlying mechanisms (Figure 12). This targets the efficient development of the application software and control strategies, leading to faster verification of the connected software functions even if the libraries were still missing.

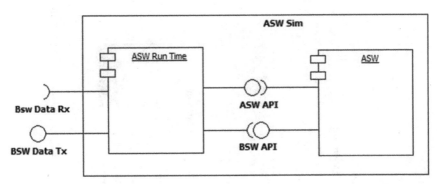

Figure 12 Simulation environment for BSW-independent ASW testing.

3. The calibration environment for the battery state estimation control strategies targets the capability to efficiently adapt the proposed algorithms to different chemistries, and the capability to optimize the algorithms to specific detailed physical behaviours for a given chemistry (Figure 13).

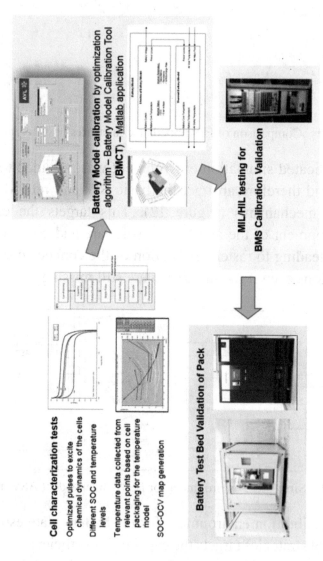

Figure 13   Depiction of the calibration and validation process of the battery core functions.

3.3 Software Platform

To enable a broad support base for the complete software development of the INCOBAT project, including the integration and the validation of the control strategies (as described above), a common software build environment has been established (Figure 14). This architecture was bounded by several requirements:

- Flexible configuration of source files, include files and directories for building code for each core;
- Sufficient intellectual property (IP) protection, for different company integrating their IP into a common computing platform;
- Adaptations to different compilers;
- Integration of additional tools;
- Minimization of the licensing costs.

A lot of work was spent on the encapsulation of all differences of the used compilers. In fact, not all partners were using the same tools for the AURIX code generation, but very different build scripts, different intrinsic functions and different linkers. The work led to the encapsulation of the variances into one header file that allows the type of the compiler to be defined, and then defines all the dependent syntax and build scripts for each tool chain. This permitted a common set of C code to be used independently from the type of compiler used.

The build process was set up to generate a separate binary image for each core. This allowed SW updates on one core without the need of rebuilding the other cores. Of course this mechanism is only applicable if the applied changes do not affect the other's core SW. The SW code allocation to the different cores was done statically through one manual configurable make file, in which for each core application source and include files or directories, pre-compiled objects and libraries can be set up.

The memory allocation was done in alignment with the AUTOSAR memory mapping approach and configured memory sections in the linker script. Depending on the currently identified CPUx in generation, the linker performs allocation of code and data to predefined flash and core local data scratchpad RAM memory sections.

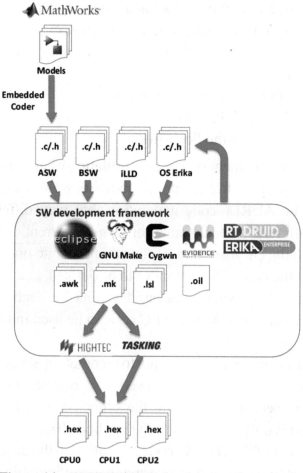

Figure 14   INCOBAT SW development framework.

The main target was the definition of a layered SW architecture (Figure 15) with well-structured interfaces, and provision of complex device drivers for the iBMS-CCU. In particular, this has been achieved through:

- Enhancement of the implementation of specific low level drivers.
- Debugging and integration of the drivers, by means of an ASW – BSW wrapper, in order to efficiently map the complex drivers to existing ASW modules, and in general follow the AUTOSAR philosophy of separation of concerns between BSW and ASW by means of well-defined interfaces.
- Safety and security mechanisms and how they have been integrated.

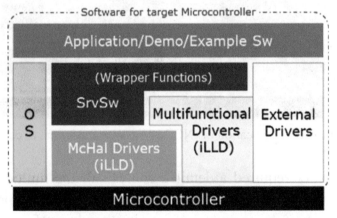

Figure 15 INCOBAT-layered SW architecture.
Source: Infineon Technologies AG.

The application software functions have been assigned to each core, paying particular attention to the safety aspects of the underlying architecture and exploiting the built-in features of the AURIX^{TM} microcontroller. Therefore, the application parts that are identified as safety relevant are assigned to core 0 and core 1, which are running in lockstep; at the same time the core 2 was reserved for the EIS that was executed only in a lab environment, to allow for sufficient computing performance.

Figure 16 below depicts the allocation of the software components to the cores, and in particular:

- An instance of the operating system ERIKA is running on each core.
- Core 0 is managing the BSW and drivers.
- The battery state estimation algorithm is run on Core 1.
- Core 2 is reserved for EIS.

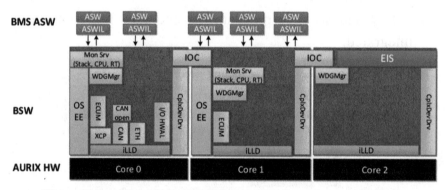

Figure 16   Allocation of INCOBAT SW components to cores.

This work required a detailed analysis of the multicore features of the architecture and the operating system support. In particular, the following decisions regarding the SW architecture were taken:

- For the multicore capabilities several SW functionalities were used similar to the currently defined and supported AUTOSAR concepts.
- Synchronized master slave start-up and shutdown approach of the cores should be used. During the start-up of master

core 0, the other two slave cores start up with a synchronization barrier during the start-up of the operating system. During shutdown the reversed order is used and master core waits until synchronized shutdown of slave cores.

- Functional based inter-core data exchange of single signals or groups similar to the AUTOSAR Inter-OS-Applicator communicator (IOC).
- Usage of spinlocks to guarantee data consistency for core to core data exchange. The spinlock mechanism was combined with immediate suspension of interrupts in order to reduce the time of remote blocking. Additionally, to prevent from deadlocks a nested acquisition of spinlocks was avoided.

Finally, in the course of INCOBAT several tests have been performed to validate the developed software architecture and the SW–SW integration aspects:

- White-box testing: verification of single SW functions such as control strategy (e.g., battery state estimation), safety function (e.g., control of battery's main relays) or basic SW (e.g., low-level drivers). Target was to provide the direct environment for these functions, therefore, sometimes shortening the SW system by investigating only one function.
- Grey-box testing: validation of SW system and especially correct integration of the functions into the control system as well as correctness of the interfaces.
- Black-box testing: validation of the safety mechanisms – especially ensuring correct reaction of the control system in case of hazardous situations.

In the context of INCOBAT, different approaches were used:

1. Model or SW in the loop (MiL/SIL): direct verification of single SW function.

2. Hardware in the loop (HiL): verification and validation of set of functions up to SW system in a real control system.
3. Vehicle demonstrator: prototyping validation in vehicle.

4 Battery Management Technology and Hardware (WP3)

The target of WP3 was to provide the required sensing and computing hardware for the INCOBAT BMS according to the requirements and specifications defined in WP1. In order to achieve the major project goals (cost reduction, industrialization, etc.), the hardware aimed to demonstrate an efficient solution in terms of cost and functionality.

Typical BMS electronic hardware architectures consist of two main parts: (i) the satellite units close to the cells or modules for cell monitoring and direct control; and (ii) the central control platform (CCU) gathering the information and managing the battery. In INCOBAT, a specific hardware for the innovative battery state estimation by EIS was implemented as (iii) daughterboard, which can be plugged to the CCU. This risk mitigation measure was taken because of the high level of innovation of the EIS. For industrialization, it would be integrated directly on the main board. The main objectives of the mentioned three hardware modules were:

i) Battery monitoring and controlling unit (satellite unit):

- Small and flexible battery monitoring unit PCB which shall be useable as slave and with slight adaptions as master.
- Outstanding EMC performance of the battery monitoring unit: DPI up to the power of 37 dBm, BCI up to at least 200 mA.

- Minimized passive balancing self-heating by the balancing resistors.

ii) Central control platform:

- Close to production by innovative, close to market automotive suitable components like safety power supply and multicore controller.
- Cost effective due to high integration and high functional density.
- Support of functional safety up to ASIL-D.

iii) EIS daughterboard:

- High precision stimulus generation and response acquisition for EIS measurements.

4.1 BMS Satellite Technology Development

The satellite units are usually tightly integrated into the battery modules and are subject to harsh environment. Important related requirements are the following:

- Scalable battery management solution for battery packs up to 1000 V
- Monitoring of voltages, temperatures, and current
- Over-/under-voltage detection
- SoC, SOH, SoF calculation
- Cell balancing: active/passive
- Robust design against RF disturbances (robust communication)
- Galvanic isolation on all major interfaces
- Fault detection and diagnostic
- Safety and (cell) protection
- Low cost
- High reliability/robustness

The core of the satellite unit is the TCB31 chip. It is a multi-cellular battery monitoring and balancing system chip designed for Li-Ion battery packs used in hybrid electric vehicles, plug-in hybrid electric vehicles and battery electric vehicles. The TCB31 is capable of monitoring the voltage and temperature of up to 12 battery cells. Up to 64 TCB31 devices can be connected by a differential serial bus interface to minimize wiring effort on the PCB. The TCB31 provides high design flexibility by supporting passive as well as active cell balancing. In the scope of INCOBAT, the communication interface was improved to become more robust against EMC disturbances and to reduce own emissions. The communication interface is used for communication among the satellite units and also for the communication to the central platform in INCOBAT. Furthermore, the measurement accuracy over the operating range was improved. The below mentioned safety features of the TCB31 were considered during the INCOBAT safety requirements collection, analysis and safety measurement implementation:

- Two independent internal voltage references;
- Secondary Monitoring of Each Cell by a 10 Bit SAR ADC;
- CRC Error Detection and Watchdog for IBCB Communication and Internal Registers;
- CRC and Parity Bits for Detection of SPI communication Errors;
- Fault Output Pin.

The highlights for the INCOBAT satellite boards relate to the TCB31 (Figure 17). These are the very robust communication interface which can be operated without transformers, the simultaneous cell voltage measurement of all battery cells with an accuracy of ± 1.5 mV and a highly efficient active balancing mode for inter-cell balancing as well as for inter-block balancing. For INCOBAT it was decided to go for passive balancing. In the course of the project, filter components were optimized for the communication interface and the voltage and temperature sense lines. Additionally, a voltage pre-regulation was put onto the board as a result of the activities which are described in the following.

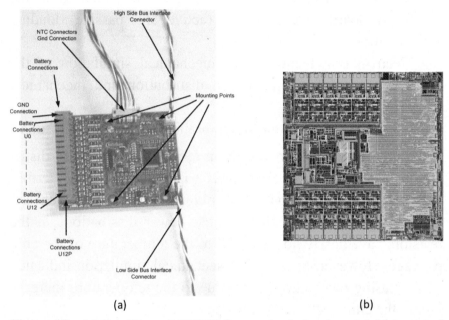

(a)   (b)

Figure 17 (a) Satellite unit TCB31 for passive balancing and (b) layout of the chip.

The developed BALI board was also subject to intensive thermal and thermo-mechanical stress investigations. Since the balancing resistors of the BALI board are used to dissipate energy from cells having a higher state-of-charge level than other cells in order to maximise the useable battery capacity, a local temperature increase at these resistors were induced, which in turn caused thermo-mechanical stress on the board and thus on the chip package. The resulting mechanical deformation of the measurement chip in turn affects the accuracy of the cell voltage measurement due to the fact, that this mechanical stress has direct impact on the bandgap voltage reference (value) being used for the ADC. For this reason, various investigations were performed, including:

- Experimental characterization of thermal behaviour of the BALI board under various (active and passive) loading conditions.
- Electro-thermal and thermo-mechanical simulation of the BALI board for temperature distribution and mechanical stress evaluation.
- Warpage measurement for calibration of simulation results.

In Figure 18, simulation results in terms of temperature distribution over board and chip surface (Figure 18a) and thermo-mechanical stress distribution over chip surface (Figure 18b) are depicted. These results showed that power dissipations in the resistors have nearly no influence to the temperature of the chip package. However, the thermo-mechanical simulation indicated an increasing mechanical stress due to the temperature increase within the chip package during operation.

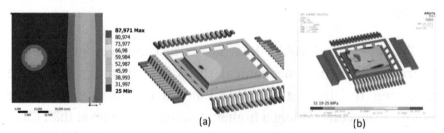

Figure 18   (a) Simulated temperature distribution over board and chip surface during operation (power dissipation of 12×80 mW in the balancing resistors and of 2.75 W within the measurement chip); (b) Simulated mechanical stress distribution over chip surface.

Experimental warpage measurements showed a difference of 2 μm for a measured temperature swing of 26 K, which was confirmed by simulation, where a deformation of 1.6 μm above the chip's hot spot was observed (Figure 19).

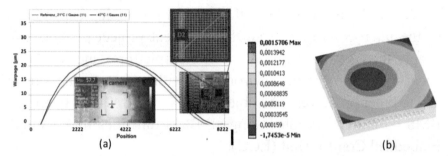

Figure 19 Measured (a) and simulated (b) warpage over chip surface during operation condition.

Consequently, a redesign of the BALI board was triggered in order to remove the heat source (voltage regulator) from the measurement chip and by this to improve the systems accuracy and robustness.

Further information can be found in "Thermo-Mechanical Stress Investigations on Newly Developed Passive Balancing Board for Battery Management Systems." (Table 1, Item 18).

4.2 Central Platform Development

One goal within INCOBAT was to realize a 'close to production' BMS demonstrator platform to enable the various partners to perform research on BMS architectures and the Electrochemical-Impedance Spectroscopy. The benefit of using the latest automotive components in a research platform is that the time taken for the step from research to a production ready solution can be shortened, as techniques employed during the research can be more directly applied. Added to this is the ever greater affordable computational power of modern multicore microcontrollers such as the new AURIX^{TM} TC275T from Infineon. The integration of many BMS functions, with many different requirements into a single CCU was possible as the parallel processing capabilities of the microcontroller mean that different project partners can supply their software into one common platform and run in an encapsulated processing environment, without interferences from and to other processes. The objective in INCOBAT was to reduce BMS costs by fully integrating all BMS functions into a single Embedded Control Unit (ECU).

The CCU is novel in many ways due to the high integration and high functional density. It also supports functional safety up to ASIL D so supports even the highest degree of rigor required for mission critical processing. The design integrates the HV-monitoring circuits using an area of the PCB which was galvanically isolated from the rest, again saving cost, reducing components and overhead and increasing performance and reliability. Due to the need for digital signal generation and analogue voltage measurement in the HV domain, standalone devices were integrated and galvanically isolated by digital interfaces over SPI.

The hardware was supplied in a metal housing made of aluminum. This allowed a fully grounded base plate to be used around the undersides of the PCB as well as a removable lid for prototyping and research. An additional plastic cover was also included inside the top metal cover to protect engineers from accidental electric shock when working on the CCU or demonstrating it to others. The connectors and mating halves were sealed, so the completed case is watertight to IP55 (Figure 20).

Figure 20 INCOBAT BMS CCU prototype hardware (EIS daughterboard not mounted).

The INCOBAT BMS CCU is based on the Infineon multicore processor AURIX^{TM} TC275. This device supports the concurrent execution of mixed ASIL functions up to ASIL-D. It offers a rich set of peripherals such as A/D converters for data capturing and it has a reasonable number of IOs to support BMS applications. In conjunction with the specific power supply ASIC TLE35584 it is possible to supply the CCU and support ISO 26262 requirements with a minimum number of components.

There is a particular HV area, HV Interlock/ADC, which is shown in Figure 20, leftmost picture, upper right part of the PCB. This connects to the HV battery and enables the integration of functions which are usually provided by a separate HV control board. Of course, the CCU design considers HV requirements which apply particular to this area like insulation and creepage distances. For safety reasons, there is a specific plastic cover for this area of the CCU (not shown in Figure 20). The ADC channels provided by this HV area are intended to be used for the novel and innovative EIS battery state estimation approach.

Low- and high-side drives are available to control contactors for various components such as DC/DC charger. Several digital inputs and low voltage ADC inputs are available. Well known state of the art communication interfaces like CAN-FD, USB and 100BaseT Ethernet are available. For early technology adoption and exploration a novel BroadR-Reach (BrdR) transceiver is provided. An SD card slot is available to record historical data and to support (software) development.

The configuration/buffers area provides a real-time clock with periodic alarm and time of day measurement. An accelerometer is included here for crash detection. Finally, the BMS System ICs provide a communication channel to Infineon's advanced Battery Monitoring and Balancing IC for automotive and industrial applications. At the edges of this area there are two connectors (horizontal bars in Figure 20, leftmost picture, upper left part of the PCB). These connect to a daughterboard which provides specific functions for the EIS feature, mainly the generation of the stimulus signal. Of course this approach contradicts the cost objective. The reason why this is not implemented on the main board is the increased flexibility to support changes of the hardware.

4.3 EIS Daughterboard Development

The EIS daughterboard integrates all the analogue circuitry needed to interface a 12 cells battery module, amplifies the cells' EIS signals and provides them, after the proper signal conditioning, to the AURIX^{TM} microcontroller on the iBMS-CCU for acquisition and processing. The daughterboard includes a voltage DAC as well, in order to provide the proper EIS current command signal to the external power transconductance amplifier, which should drive the EIS current stimulus in the battery module under test.

The EIS daughterboard (Figure 21) features the following functions:

- Opto-isolated switches to disconnect the battery cells when measurements are not performed.
- EIS Command Generator (DAC, voltage output).
- 12× EIS cell voltage measurement circuitry, composed by differential amplifier, OCV cancelling circuitry, and 4th order Bessel anti-aliasing filter.
- 2× EIS current measurement circuitry, to provide the same signal conditioning (4th order Bessel anti-aliasing filter) of the EIS cell voltage channels to the EIS current signals.

Figure 21 EIS daughterboard plugged on iBMS-CCU.

At this point, an EIS test bench has been set up for func-·tional validation (both HW and SW), including the iBMS-CCU equipped with the EIS daughter board, an adequate DC power supply and the EIS current amplifier, to measure the impedance of a single battery cell under test. Due to the limitations of the EIS current amplifier, intended for the validation of the EIS measurement method and not the in-vehicle acquisition of a complete battery pack, only one battery cell at a time could be tested. The setup used for EIS functional validation is shown in Figure 22, while the EIS measurements results for a known new cell (EIG EPLBC020B, serial number 9E28A-086) provided by the iBMS-CCU and the impedance measurements obtained via lab equipment are shown for comparison on the same graph in Figure 23 below.

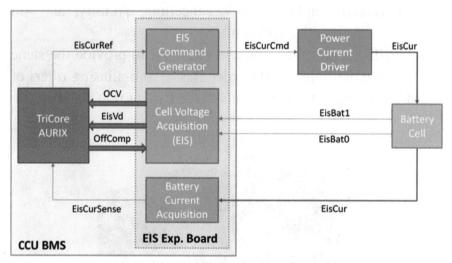

Figure 22   EIS functional validation setup.

A good correlation can be noticed between EIS results provided by the iBMS-CCU and the lab measurement, apart from the low frequency range where some oscillation is present in the EIS phase measurements. This effect was caused by a problem in the measurement sequence that was adopted and subsequently solved: in fact all electrical transients require some complete periods to stabilize, while the EIS method requires a steady state condition. Therefore, the issue can be avoided by simply starting the EIS signal acquisition after some (at least two) complete periods at the lowest frequency of the EIS signal (0.01 Hz or 100 s).

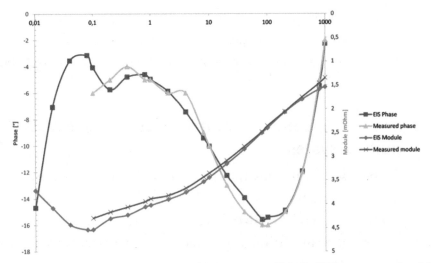

Figure 23 Results of EIS algorithm run on iBMS-CCU compared with measured values with lab equipment.

After both HW and SW for the EIS measurements had been validated, the iBMS-CCU equipped with the EIS daughterboard has been used to successfully investigate how the battery cell impedance varies as a function of the cell health, measuring a set of new, used and abused battery cells provided by Manz. The EIS current amplifier has been used to force the EIS current signal into the battery cells under test.

5 Integration and Verification (WP4)

The target for WP4 was the integration and verification of the proposed BMS platform into a high-voltage (HV) battery system in a first step and in a second step into a demonstrator vehicle. In this work package inputs from WP2 (software platforms, control strategies) and WP3 (hardware platform) and are merged into an integrated control system, tailored a defined HV battery system and in the end, integrated into a vehicle demonstrator.

Consequently, a proper management of heterogeneous skills was required for this work package (e.g., mechanical design, vehicle engineering, HW and SW design, and knowledge about the related standards).

5.1 Recommendations for Integration of the INCOBAT Battery System

The list below summarizes the integration recommendations for the INCOBAT battery system and thus provides a basis overview for the integration of the BMS platform into the vehicle. During the development the following standards have been taken into consideration[1]:

- ISO 26262 – "Road Vehicles – Functional Safety"
- ECE R100 – "Approval of Battery Electric Vehicles with regard to specific requirements for the construction, functional safety, and hydrogen emission"
- SAE J1766 – "Recommended Practice for Electric and Hybrid Electric Vehicle Battery Systems Crash Integrity Testing"
- ISO 6469.1 – "Electrically propelled road vehicles – Safety specifications. On-board rechargeable energy storage system (RESS)"
- FMVSS 305 – "Electric powered vehicles: electrolyte spillage and electrical shock protection"
- ECE R10 – "Uniform Provisions Concerning the Approval of Vehicles with regard to Electromagnetic Compatibility"

---

[1]Since the INCOBAT project was a research project, full standard compliance and series-production readiness was not intended; consequently, additional efforts will be required for future series production projects.

- CISPR 25 – "Radio disturbance characteristics for the protection of receivers used on board vehicles, boats, and on devices – Limits and methods of measurement"
- ISO 10605 – "Road vehicles – Test methods for electrical disturbances from electrostatic discharge"
- ISO 16750 – "Road vehicles – Environmental conditions and electrical testing for electrical and electronic equipment"
- ISO 7637 – "Road vehicles – Electrical disturbances from conduction and coupling"
- SAE J1979 – "E/E Diagnostic Test Modes"
- SAE J2012 – "Diagnostic Trouble Code Definitions"
- SAE J2178 – "Class B Data Communication Network Messages"
- ISO 17987 – "Road vehicles – Local interconnect network (LIN)"
- ISO 11898 – "Road vehicles – Controller area network (CAN)"
- IEEE 802.1-3 – "Ethernet 10BASE-T/100BASE-TX/ 1000BASE-TX over twisted pair"
- IEEE 802.3-2012 – "BroadR-Reach ethernet physical layer"

5.2 Battery Pack Design

The mechanical integration of the battery pack into an existing vehicle chassis induced some additional constraints. The housing had to fit into the vehicle spacing in a way that the battery pack housing is mechanical stiffness and bear any driving forces, considering that it is not recommended to cut any of the original elements of the main frame or main plate. Nevertheless, the construction battery pack housing is prepared for road conditions and does not require additional covering or sealing. For attaching the battery pack in the chassis, additional protective covers for preventing the battery cell housing from possible stone hits during the ride is recommended. Special attention must also be paid to proper bending angles of the power cables and possible changes of the cable material (stiffness especially at the endings) over time, since these points are the most critical in terms of mechanical shocks.

The INCOBAT battery block has been split into two parts: the front part (depicted in Figure 24) consists of 12S2P cells and all necessary protection elements and the rear one (depicted in Figure 25) consists of 24S2P cells and basic protection measures. The front battery pack is designed for an enclosed environment (such as trunk or under-bonnet area) and the rear batter pack for installation below the car body (outside environment).

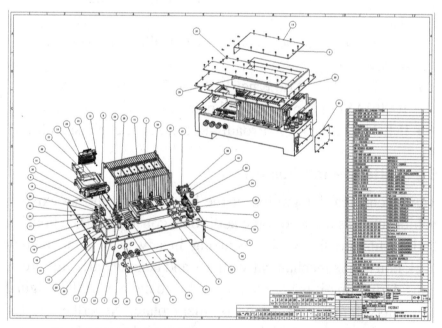

Figure 24 INCOBAT front battery pack (exploded view).

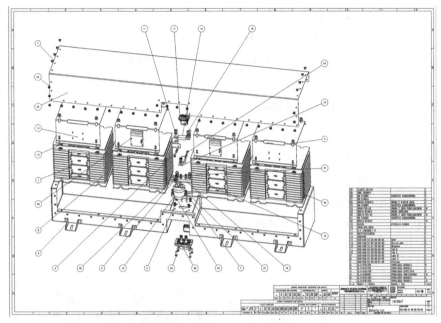

Figure 25 INCOBAT rear battery pack (exploded view).

5.3 INCOBAT Battery System

The battery pack was configured/designed for following specification:

- Connection structure: 36S4P
- Nominal HV voltage 135 V
- Maximum HV voltage 150 V
- Peak load (10 s) >35 kW
- Maximum peak current 200 A
- Continuous discharge current 100 A
- Battery pack capacity 14.2 kWh
- Charge discharge temperature range 0 to +40°C
- Discharge temperature range –20 to +40°C
- Battery lifetime 10 years or 200,000 km
- Battery charge/discharge cycles >1500
- Cell in use Enerdel CE210

- Power cable connection Omerin ECS silcable 25 mm^2
- Power contactor Gigavac EV200BAB
- Power connector Delphi RCS890 9-2141227-1

5.4 Vehicle Integration

The vehicle demonstrator platform used for INCOBAT was a Renault Twingo. Necessary adaptions to the vehicle included removing of fuel installation, engine, and exhaust installation, subsystems like vacuum break pump, power steering pump, cabin heating, and motor cooling and the respective replacement by electric driven equivalents.

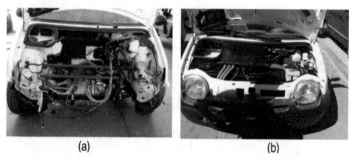

(a)   (b)

Figure 26   (a) Vehicle front partly disassembled, with battery pack mounting frame and (b) front battery pack integrated.

In order to facilitate the modifications performed on the under-bonnet area, the front of the vehicle has been dismantled (as shown in Figure 26) and the engine was replaced by a new electric motor with a customized gearbox with a differential mechanism box. As a high packaging density of components in front of the vehicle is very common, the rework of the vehicle allows to open the front to easily access all the elements which are subject to modification.

For safety reasons the most important element to replace in the car is the vacuum pump, normally driven by shaft of the combustion engine, used to generate a vacuum necessary for the work of many systems, e.g. servo brake, any type of EGR valves, N75, waste gate, etc. In the case of an electric vehicle solely, the support for brake pressure regulator is required. An electric car does not have EGR valves or other vacuum controlled apparatus, but the brakes remain hydraulic. In a hydraulic braking system, the vacuum is necessary for the servo pump – since after breaking the pedal is released vacuum retracts the brake pads. In case of electric vehicle this auxiliary element must be replaced by an electrical powered version.

Figure 27 Electric motor in the frame, attached to gear/shaft box (a); gear/shaft box with different splines visible (b).

In Figure 27, panel a shows the installed electric motor, a three-phase BLDC motor, mounted to the frame of the car. The gearbox and differential are directly mounted to the motor shaft. In case of the electric car, constant control of gear is not necessary, as the parameters of the direction of rotation, torque power and speed can be controlled electrically. Very often, when converting the internal combustion car to an electric vehicle, the original gearbox designed for the car in locked position is used. This usually results in sufficient compromise between range of car speed and torque. Nevertheless, the effectiveness of the gear-box of the combustion engine car in relation to the transmission adapted to operate in the electric car was not analysed in this project.

5.5 Modification in the Engine Control Unit (ECU)

The vehicle ECU firmware has adapted to the changes introduced in INCOBAT to the vehicle CAN interface protocol. New signals and required corresponding functionalities were added. The main areas of the ECU modifications were:

- The sourced power reduction – the maximum motor RPM are BP SOC dependent. The power sourced from the BP can also be reduced/turned off per CCU (BP) request.
- The CCU communication loss detection, and
- Overall interface compatibility.

In current vehicles, communication between internal subsystems is essential for correct operation. There are less important components whose presence do not affect the driving process itself (controllers wipers and window lifting), while other elements associated with the drive are necessary (inverter, battery). This is particularly important in the case of an electric car where the battery has to be checked constantly for safety reasons. Loss of important information about the status of cells can lead to damage or even worse – a fire on board of the car. In case of loss of communication with key components of the car, the battery must perform an emergency system shutdown. The integration of these elements as well as examining their proper operation should be performed primarily place during the process modifications of a car. Figure 28 illustrates the adapted main state machine algorithm of the ECU.

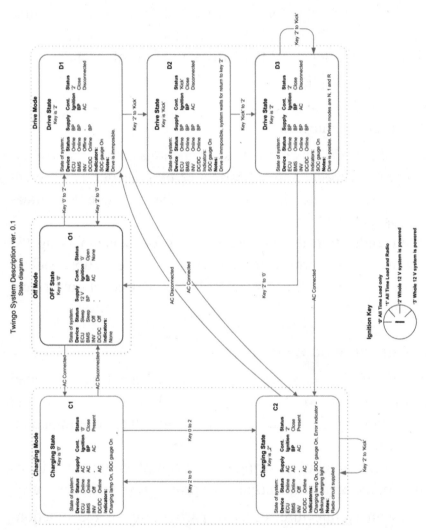

Figure 28 Main machine state algorithm of the ECU.

6 Platform Evaluation (WP5)

Verification and validation of the developed INCOBAT system played an essential part in allowing a further commercialization of the developed technologies and methodologies after the course of the project. Therefore, the main objective of WP5 was to evaluate the developed BMS platform in terms of

- ICT functionalities,
- industrialization aspects,
- and vehicle integration.

In case of the ICT functionalities, a benchmarking of the demonstrator with respect to its performance by considering different aspects, such as efficiency, accuracy of battery state determination, was realized.

Industrialization aspects are focusing on benchmarking the demonstrator platform with respect to cost reduction, safety/security as well as reliability and lifetime assessment of critical components.

6.1 Performance Evaluation

The aim of this task was the accomplishment of dedicated performance tests to validate the targeted improvements of the BMS regarding energy efficiency, battery state determination, and system complexity.

The main target of reducing system complexity was the focus in most of the enhancements and the project results; in particular:

- System architecture: efficient partitioning of the functionalities, impact on development efficiency; increasing the consistency within the development between the skills, and over project lifetime.

- Integration of multiple functionalities: CPU Performance from AURIX^{TM} (TC27xx) and opportunity for ASW integration (reduction of number of xCU).
- TriCore AURIX^{TM} platform: Cost of CPU (65 vs. 40 nm).
- Smart and integrated module management units: cost of satellite systems (IBCB communication, reduced wiring, higher accuracy over temperature range and lifetime).
- Modular SW platform: License costs (industrial solution with license fee vs. freeware solution Erika).
- Improved BMS control algorithms (EIS and advanced Kalmann): better assessment of SoH (lower aging effects and higher lifetime).
- Reliability and robustness assessment: costs for V&V, impact on design for diagnostics/maintenance.

Further, a target of the task was to state the energy efficiency of the battery state algorithms. In this aspect several tests were performed to validate the results of the

- Accuracy of SOC estimation,
- Functionality of SOF estimation, and
- Current demand is limited by the SOF estimates in order to validate the limit calculation.

This aspect targets the capability to efficiently adapt the proposed algorithms to different chemistries and the capability to optimize the algorithms to specific detailed physical behaviors for a given chemistry.

6.2 Reliability/Lifetime Testing and Validation

The consideration of non-functional aspects such as reliability and robustness was an essential part of the project for paving the way to future industrialization. For this reason, methodical developments and dedicated reliability assessment measures were conducted, including:

- Development of a full-field optical technique based on focus-stacking to measure temperature dependent warpage and in-plane deformation.
- Investigations of new reliability test schemes based on applying multiple load factors.
- Thermo-mechanical robustness analysis, which has been focused exemplary on the satellite BALI board in order to identify potential weak points (and to derive design optimization guidelines, if required) and to validate the developed and above-mentioned methodologies.

Within the INCOBAT project, CWM GmbH continued developing a method, which, in addition to the in-plane deformation measurement, takes the out-of-plane deformation behavior of, e.g., PCB boards during thermal loading into account. A prerequisite for a perfect in-plane measurement is that the surface to be measured always remains flat during the load. If this is not the case, a measurement error which distorts the results is generated. In order to solve the problem, it is necessary to consider the out-of-plane component in addition to the in-plane displacement.

In general, a variety of various measurement methods for determining the out-of-plane component exist. In order to avoid additional effort and costs, the use of the same measurement principle, which is also used for the in-plane measurement, was targeted. These considerations result in the development of a full-field optical technique based on focus-stacking in order to measure warpage caused by external load (Figure 5). Compared to other full-field optical measurement techniques this method has the following advantages:

- In-plane deformation and strain measurement are possible.
- Out-of-plane warpage measurement at the same load state.
- Lateral resolution is only limited by the image-recording technique.
- Simple hardware configuration.
- (Almost) no specimen preparation.
- Scalable from board level down to component level.

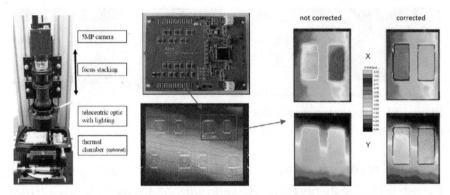

Figure 29 Correction of the in-plane displacement of SMD resistors by out-of-plane component during thermal load.

A technology readiness level of 8 (system complete and qualified) was achieved to obtain the result of these developments during the project runtime. The commercial system is marketed under the name "microDAC® profile" by CWM GmbH (Figure 29, left).

A further topic was the investigations of new multiple-load reliability tests schemes. The basic idea by this was to stimulate more realistic failure mechanisms and at the same time to reduce the overall testing period by applying multiple load factor patterns, including for example thermal cycles, vibration and humidity.

In the first step, various investigations were performed on own test boards containing dedicated daisy chain structures in order to enable a mapping of the different daisy-chain/component types with respect to the applied load scenario, achieved number of cycles to failure as well as the underlying failure mechanism (e.g., location and nature of solder joint cracks, Figure 30).

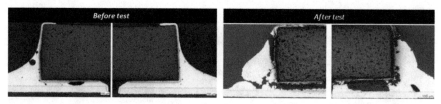

Figure 30   Exemplary cross sections of SMD 0603 resistor before and after reliability testing.

In the second step, these investigations were used as baseline for reliability tests performed exemplary on the within INCOBAT developed BALI board. This electronic board was identified to be suitable for this purpose, since it is more prone to reliability risks than the remaining BMS boards due to the additional heat dissipation at the balancing resistors. Furthermore, there was already a good knowledge base available regarding its thermal and thermo-mechanical behavior due to the previous studies. The test setup for combined testing (boards are placed inside a temperature chamber and attached to a shaker simultaneously) together with an exemplary temperature/humidity profile is shown in Figure 31.

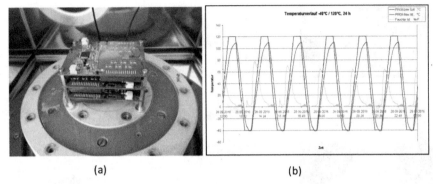

(a)   (b)

Figure 31 Test setup for multiple-load test (a: BALI boards attached on shaker and b: Temperature/humidity test profile).

An exemplary cross section of a tested (thermal shock test: $-40/125°C$, $t_{dwell} = 20$ min) balancing resistor that has run 1200 cycles is shown in Figure 32a. In this particular case, delamination between device and solder joint could be observed for both contact areas.

Furthermore, a FE analysis was performed in order to assess the thermo-mechanical strength of the mounted SMD balancing resistors and to predict the lifetime (Figure 31b). This analysis predicted a beginning of the solder joint cracking (for the SMD resistors) after approx. 800 temperature cycles and a complete cracking after approx. 1500 temperature cycles, which thus was well in line with the experimental test results.

As a conclusion, it can be stated that the investigated BALI board is, according to the FE analysis and tests results, fully achieving the lifetime specification with respect to temperature cycling testing (1000 cycles, as required for automotive applications). Furthermore, using the calibrated and validated FE model, further design variations can be investigated without employing long-lasting experimental tests in order to optimize the reliability of the system (= virtual prototyping) in the future.

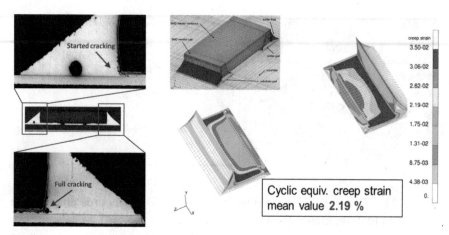

Figure 32   Tested balancing resistor after 1200 cycles with started and completed delamination in the contact areas (a) and corresponding FE geometry analysis results (b).

6.3 Safety/Security Testing and Validation

The target of this task was the validation of safety relevant aspects. This work mainly relied on two important inputs: hazard and risk analysis as well as FMEA performed in WP1 in order to identify the main risks for the system in a systematic way, and the mini HiL test environment developed conjointly with WP2 and WP4 for the integration of the SW in the control system and the validation of the control strategy.

Based on the findings of the FMEA, a comprehensive set of countermeasures was defined and implemented (at least one countermeasure for each safety goal and security target was ensured). The conceptual depiction of this mapping of FMEA to counter measures is shown in Figure 33 and results in the functional safety concept (FSC) of the INCOBAT system.

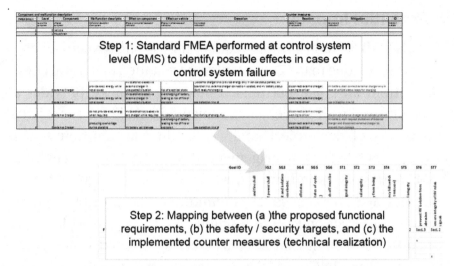

Figure 33 Conceptual depiction of FMEA to counter measure mapping.

The validation focused on the different development levels. The development and validation plan (DVP) (INCOBAT Consortium, 2016) provided guidelines for the different abstraction levels and components and also included relevant measures and test specifications.

Objective of the safety verification and validation activities were the provisioning of evidence of compliance with the safety concepts and the correct, complete and consistent implementation of safety measures. Evidences of the various test campaigns can be found in the 'Report on encryption technologies for BMS safety and security' (deliverable D3.3 (INCOBAT Consortium, 2014)) and D1.4 (INCOBAT Consortium, 2016) as well as deliverable D2.5 (INCOBAT Consortium, 2016).

7 Potential Impact

The INCOBAT consortium, combining strong expertise and diverse nature due to the consistent representation of several companies from different, but related fields of business, provides a successful and well proven environment to enable the successful completion of the project goal – providing an innovative and cost efficient battery management system for next generation HV-batteries. Coming from different industrial backgrounds and countries and showing great commitment, the project partners are able to target and serve a unique combination of different networks throughout Europe, thus creating a sound exploitation platform for the INCOBAT project results and outcomes.

The impact of the INCOBAT project has been even more strengthened by INCOBAT's participation in the Cluster "4th Generation EV Vehicle". Within the cluster, INCOBAT was active in three working groups, each focusing on different topics, trying to ensure the sustainability of project outcomes of all partner projects.

Exploitable results have in fact been identified in the very early stages of the project for each project partner, in order to be able to closely monitor the development and to take possible required protection measures.

The baseline for the exploitation of the INCOBAT project results were the envisaged technology development targets – either product and IP development (e.g., AURIX^{TM}, BALI, and EIS) or engineering services (e.g., safety and security, battery assembly, and reliability).

For the first category, IP protection and product management was performed on company level. For the second category, the path for exploitation focuses on intensive dissemination of technology in order to make potential customers aware of the partner's competences. The INCOBAT project is therefore used as basis to apply the developed methods in a realistic context.

7.1 Exploitation of Results

In order to successfully exploit project results, the INCOBAT consortium developed a sustainability strategy, outlining what should happen to the project outputs at the end of the project, exploring how they are maintained after project funding has stopped. Sustainability of results implies use and exploitation of results in the long term. A clear and target-oriented dissemination and exploitation strategy will enhance the impacts of the project by promoting the project and its findings to the relevant audiences therewith to achieve the largest possible impact of the project.

In INCOBAT, the sustainability model was built around three main focus points:

BENEFIT: What are the long-term benefits (technical and non-technical results and outcomes) that the beneficiary gained in the project?

TARGET: Who is a possible target for the exploitation, who are the stakeholders? Which networks are interested in the project outcomes?

MEASURE: Which measures should be taken in order to maintain the benefit for a certain target group?

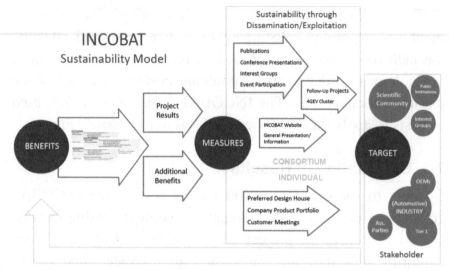

Figure 34   INCOBAT sustainability model.

The project outcomes have been assessed individually by each beneficiary as well as jointly by the whole consortium. Certain project results have been defined as beneficial to the individual project partners – be it planned project results (like for example the BALI Chip (TCB31)/interchip connection bus, the EIS and the microDAC profile) or welcome side-effects (like cooperations with large partners leading to higher visibility of I the market).

The measures for exploitation are divided in two groups – measures that are taken on a project level, involving various (and in some cases all) consortium members as well as individual exploitations measures. The first group usually targets a broader audience, including the scientific community and the general public, while the second is focusing on existing and possible future customers.

A tailoring to the specific Stakeholder groups is of utmost importance, not only in order to ensure the actual commercial use of project results, but also to use the stakeholder influence to in return ensure that project benefits stay what they are – benefits. In the sustainability model this is indicated by a backwards arrow from Stakeholders, showing how through external influences the project benefits are subject to change.

On a consortium level, INCOBAT is represented on the project website[2], that has been continuously maintained by the coordinator and serves as a hub for many of the ongoing dissemination activities. On the website, the following dissemination material is publically accessible:

- general project information,
- public summaries of deliverables, and the
- public summary report.

The opportunity to contact the project consortium (through the project coordinator) will furthermore also be available after the project. This ensures the possibility of accessing the project results even long after the project ended and enables possible customers or interested parties to get in contact with individual project partners.

The sustainability of project results will additionally be strengthened by the participation in interest groups on a European level and the interaction with international partnerships and counterparts (amongst others through EUCAR, ACEA, ERTRAC, EARPA), in order to communicate and disseminate the knowledge gained within the project to the international transport community and beyond.

---

[2]www.incobat-project.com

On an individual level, the project beneficiaries were highly encouraged to plan the exploitation of their respective results from a very early stage of the project until the end and especially beyond the project. Therefore, an individual exploitation plan is available for all INCOBAT beneficiaries.

7.2 Main Dissemination Activities

The dissemination for the project was – as initially planned in the project proposal – conducted on two levels; coordinator-driven and partner-driven, thus addressing different target audiences through different channels, in order to gain a high level of attention.

While the coordinator-driven dissemination aimed at representing the entire INCOBAT project, focusing on common objectives and results and the partners as a collective, the partner-driven activities are conducted by the project partners themselves and thus rely on partner initiative.

Alongside the above dissemination activities, the INCOBAT consortium issued a total of 21 peer-reviewed publications.

Table 1 INCOBAT publications

No.	Title	Conference Short Name	Main Author	Year
1	Seamless Model-Based Safety Engineering from Requirement to Implementation	MODELS	G. Macher	2014
2	Automated Generation of AUTOSAR Description File for Safety-Critical Software Architectures	ASE Workshop	G. Macher, E. Armengaud, and C. Kreiner	2014
3	Bridging Automotive Systems, Safety and Software Engineering by a Seamless Tool Chain	ERTS Conference	G. Macher, E. Armengaud, and C. Kreiner	2014
4	Automotive Real-time Operating Systems: A Model-Based Configuration Approach	EWiLi Workshop	G. Macher, M. Atas, E. Armengaud, and C. Kreiner	2014
5	SAHARA: A Security-Aware Hazard and Risk Analysis Method	DATE	Georg Macher, H. Sporer, R. Berlach, E. Armengaud, and C. Kreiner	2015
6	Improving HV battery efficiency by smart control systems	SSI	Armengaud, Macher, Kurtulus, Groppo, Haase, Hofer, Lanciotti, Otto, Schmidt, and Stankiewicz	2015
7	Integration of Heterogeneous Tools to a Seamless Automotive Toolchain	EuroSPI	G. Macher, E. Armengaud, and C. Kreiner	2015
8	Model-Based Configuration Approach for Automotive Real-Time	SAE	G. Macher, M. Atas, E. Armengaud, and C. Kreiner	2015

(Continued)

Table 1   Continued

No.	Title	Conference Short Name	Main Author	Year
9	A Combined Safety-Hazards and Security-Threat Analysis Method for Automotive Systems	SAFECOMP	G. Macher, A. Hoeller, H. Sporer, E. Armengaud, and C. Kreiner	2015
10	A Comprehensive Safety, Security, and Serviceability Assessment Method	SAFECOMP	G. Macher, A. Hoeller, H. Sporer, E. Armengaud, and C. Kreiner	2015
11	Service Deterioration Analysis (SDA): An Early Development Phase Dependability Analysis Method	RADIANCE	G. Macher, A. Hoeller, H. Sporer, E. Armengaud, and C. Kreiner	2015
12	A Versatile Approach for ISO 26262 compliant Hardware-Software	SAE	G. Macher, H. Sporer, E. Armengaud, and C. Kreiner	2015
13	Automated Generation of Basic Software Configuration of Embedded Systems	ACM RACS	G. Macher, R. Obendrauf, E. Armengaud and C. Kreiner	2015
14	A smart computing platform for dependable battery management systems	AMAA	E. Armengaud, C. Kurtulus, G. Macher, M. Novaro, G. Hofer and H. Schmidt	2015
15	Using Model-based Development for ISO 26262 aligned HSI Definition	CARS@EDCC	G. Macher, H. Sporer, E. Armengaud, E. Brenner and C. Kreiner	2015
16	A Seamless Model-Transformation between System and Software Development Tools	ERTS[2]	G. Macher, H. Sporer, E. Armengaud, E. Brenner and C. Kreiner	2016

	Title	Venue	Authors	Year
17	RTE Generation and BSW Configuration Tool-Extension for Embedded Automotive Systems	ERTS[2]	G. Macher, R. Obendrauf, E. Armengaud, E. Brenner and C. Kreiner	2016
18	Thermo-Mechanical Stress Investigations on Newly Developed Passive Balancing Board for BMS	AmE 2016	Alexander Otto, Florian Schindler-Saefkow, Sven Haase, Lutz Scheiter, Günter Hofer, Eric Armengaud and Sven Rzepka	2016
19	Embedding electrochemical impedance spectroscopy in smart battery management systems using multicore technology	AMAA	E. Armengaud, G. Macher, R. Groppo, M. Novaro, A. Otto, R. Döring, H. Schmidt, B. Kras and S. Stankiewicz	2016
20	Thermo-Mechanical and Mechanical Robustness of the INCOBAT Smart Battery Management System	SSI 2017	Alexander Otto, Ralf Döring, Lutz Scheiter, Eric Armengaud and Sven Rzepka	2016
21	Migration of automotive powertrain control strategies to multi-core computing platforms – lessons learnt on smart BMS	ERTS[2]	Eric Armengaud, Ismar Mustedanagic, Markus Dohr, Can Kurtulus, Marco Novaro, Christoph Gollrad and Georg Macher	2016

8 Main Achievements and Outlook

The main INCOBAT technical achievements can be summarized as follow:

- Improving the range of the electric vehicle by better use of the electrical energy stored within the battery, realized by a combination of TI01 (mission profiles), TI03 (efficient partitioning), TI05 (multicore computing platform), TI06 (smart module management unit) and TI08 (improved BMS control algorithms).
- Significant decrease of costs for BMS hardware, realized by a combination of TI03 (efficient partitioning), TI04 (integration of multiple functionalities within the same control unit), TI05 (multicore computing platform), TI06 (smart module management unit), TI07 (modular SW platform), and TI08 (improved BMS control algorithms).
- Provide modular concepts for efficient integration into the vehicle, realized by TI02 (model-based systems engineering), TI03 (efficient partitioning), TI05 (multicore computing platform), TI07 (modular SW platform), TI09 (safety and security co-engineering), TI10 (design and validation plan), and TI11 (reliability and robustness validation).

Achievements regarding dissemination and exploitation of the INCOBAT outcomes include 21 peer-reviewed publications and a dedicated cluster workshop to exchange information between related projects, as well as the development of a dedicated exploitation plan and sustainability model should be highlighted.

A central aspect within INCOBAT was the consistent management and mapping of TIs with the main project objectives and the respective path for exploitation and sustainability. Hence, the TIs represent a manageable set of assets, which can be efficiently developed by one partner or a group of partners within the project, and by assembling and synchronizing these assets (usually along the supply chain), the global objectives could be reached. Similarly, the management of single TIs strongly simplifies the path for exploitation at partner level. Most of the TIs could be routed to single partner and exploited according to their internal business development strategy. Consequently, the handover from R&D to industrialization and future exploitation could be performed in a smooth manner.

As outlook for further research and innovation, following topics have been identified in the course of INCOBAT project.

- Battery technologies for **higher performances** and tailored to **customer needs:**
  - New chemistries, hybrid batteries, more accurate battery state
    estimation for better usage of energy available over the lifetime (incl. 2nd life).
  - Covering full range of road transport applications.

- Efficient integration in **comprehensive energy management concept** (both at electrified drivetrain and entire vehicle level):
  - Improved energy efficiency of electrical drive-train, energy harvesting, comprehensive thermal energy management.

- Increasing **maturity of the technology** over the entire lifecycle (including second life and recycling):
  - Dependability including safety, reliability and robustness, health-monitoring of critical components, virtual

prototyping (DfR, DfT, and DfM) and improved stress-testing methods.

- **Convergence** e-mobility and autonomous-driving functions:
  - ○ Predictive energy management taking into account road profile, environment, and traffic situation.

9 General Project Information

Project Consortium:

Partner	Country
AVL List GmbH	Austria
Ideas & Motion	Italy
Fraunhofer Gesellschaft zur Förderung der angewandten Forschung E.V.	Germany
Infineon Technologies Austria AG	Austria
Infineon Technologies AG	Germany
Impact Clean Power Technology S.A.	Poland
MANZ Italy S.R.L	Italy
Chemnitzer Werkstoffmechanik GmbH	Germany

Contact Coordination:

Mr. Eric Armengaud **E-mail: eric.armengaud@avl.com**

Contact Communication:

Ms. Nadine Knopper **E-mail: nadine.knopper@avl.com**

Project Website:

www.incobat-project.eu

Project Logo:

⁺⁻incobat

Index

Partner List

The research leading to these results has received funding from the European Union's Seventh Framework Program (FP7/2007–2013) under grant agreement n° 608988

The consortium consists of the following partners

Partner	Country
AVL List GmbH	Austria
Ideas&Motion s.r.l.	Italy
Fraunhofer Gesellschaft zur Förderung der angewandten Forschung E.V.	Germany
Infineon Technologies Austria AG	Austria
Infineon Technologies AG	Germany
Impact Clean Power Technology S.A.	Poland
MANZ Italy SRL	Italy
Chemnitzer Werkstoffmechanik GmbH	Germany

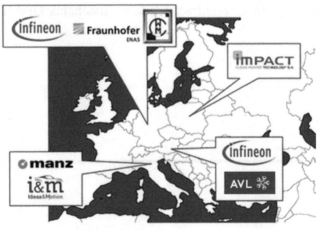

Basic map: digitale-europakarte.de, 2012-10-10

List of contributing authors

Eric Armengaud, AVL List GmbH
Nadine Knopper, AVL List GmbH
Georg Macher, AVL List GmbH
Ismar Mustedanagic, AVL List GmbH
Christoph Kreuzberger, AVL List GmbH
Markus Dohr, AVL List GmbH
Can Kurtulus, AVL Research and Engineering
Riccardo Groppo, Ideas&Motion SRL
Marco Novaro, Ideas&Motion SRL
Sven Rzepka, Fraunhofer Institute for Electronic Nano Systems
ENAS
Alexander Otto, Fraunhofer Institute for Electronic Nano Systems
ENAS
Florian Schindler-Saefkow, Fraunhofer Institute for Electronic
Nano Systems ENAS
Holger Schmidt, Infineon Technologies AG
Günter Hofer, Infineon Technologies Austria AG
Slawomir Stankiewicz, Impact Clean Power Technology S.A.
Bartek Kras, Impact Clean Power Technology S.A.
Claudio Lanciotti, Manz Italy SRL
Bettina Seiler, Chemnitzer Werkstoffmechanik GmbH
Ralf Döring, Chemnitzer Werkstoffmechanik GmbH
Kerstin Kreyßig, Chemnitzer Werkstoffmechanik GmbH

AVL www.avl.com	

AVL List GmbH is the world's largest privately owned company for development, simulation and testing technology of powertrains (hybrid, combustion engines, transmission, electric drive, batteries, and software) for passenger cars, trucks and, large engines. AVL has about 3500 employees in Graz (Austria), and a global network of 45 representations and affiliates resulting in more than 8000 employees worldwide. AVL's Powertrain Engineering division activities are focused on the research, design and development of various power-trains in the view of low fuel consumption, low emission, low noise and improved drivability. The Advanced Simulation Technologies division develops and markets the simulation methods which are necessary for the powertrain development work. The Instrumentation and Test Systems division is an established manufacturer and provider of instruments and systems for powertrain and vehicle testing including combustion diagnostic sensors, optical systems as well as complete engine, powertrain and vehicle test beds. AVL supplies advanced development and testing solutions for conventional and hybrid vehicle components and systems like simulation platforms, development tools and system integration tools.

| I&M |  Ideas&Motion |

Ideas & Motion s.r.l. (I&M), Cherasco (CN) – Italy, offers high-tech solutions in the field of automotive applications. I&M has a well-recognized position of excellence on some key areas, stemming from more than 100 years of combined experience on innovative projects and related automotive products.

The long-lasting relationships with the major worldwide silicon makers and automotive suppliers make a unique access available to the most advanced technologies in the automotive domain. The company focuses its efforts on some very clear goals:

- priority on R&D, with particular respect to the protection and exploitation of IP
- innovation transfer into products with a high technological content and limited production lots
- "fabless" organization featuring the unique access to technologies and test facilities to support the development of advanced automotive systems through its local industrial partner

Main priority on 3 application areas, as a solid basis for the subsequent expansion on other domains:

- design of integrated circuits and IPs on silicon (e.g. ASIC, ASSP)
- design, development and manufacture of automotive control systems (high tech for niche applications)
- design, development and implementation of energy management systems for hybrid/electric vehicles (FEV)

There is ambitious growth plan to train and integrate young talents in the company and the INCOBAT project perfectly fits the company's strategies.

FhG	◢ **Fraunhofer**

The Fraunhofer Research Institute for Electronic Nano Systems ENAS in Chemnitz focuses on research and development in the fields of smart system integration with partners in Germany, Europe and worldwide. It has strong experience in the design and the simulation of wired and wireless communication systems in harsh environments for a wide field of applications. Fraunhofer ENAS is a specialist of early stage feasibility studies and pre-design methodologies for customers in Automotive and Aeronautics.

With its Micro Materials Center, Fraunhofer ENAS does world-leading research on the reliability of smart systems:

- Accelerated reliability testing accounting for complex loading situations that include moisture, thermal, mechanical, electrical, diffusion, corrosion etc. effects
- Local deformation and stress measurement (microDAC, fibDAC, 3-D X-ray CT etc.)
- Advanced modeling and simulation of thermo-electro-mechanical reliability concerns in packaging and on-chip interconnects – also applying advanced fracture mechanics
- Lifetime analyses, lifetime prognosis and lifetime optimization for smart systems

| IFAT |  |

Infineon Technologies Austria AG is a legally independent subsidiary, 100% owned by Infineon Technologies AG in Germany (www.infineon.com). It is one of the globally acting manufacturing and research & development centers of Infineon Technologies AG. The headquarters of the Infineon Technologies Austria AG is located in Villach and currently employs around 2600 persons. The manufacturing facility acts as center of competence for manufacturing of power semiconductor discrete and integrated products. Around 700 engineers and researchers in Villach as well as in its subsidiaries in Graz, Linz and Vienna develop semiconductor products for the facility in Villach, as well as for all other manufacturing locations of the enterprise. Infineon Technologies Austria AG acts as a center of competence with profound system-, development-, engineering- and manufacturing intellectual property for the following business lines: Automotive (ATV), Industrial Power Control (IPC), Power Management and Multi Market (PMM), Security and Chip Card IC (CCS). Infineon forcefully follows the strategy to strengthen the R&D capability within the company. Today approximately 950 persons are employed in R&D within Infineon Austria forming one of the biggest development teams for microelectronics in Austria. In FY 2010/2011 approximately 16% of the revenue has been invested in research and development activities. The Development Center Graz is on one hand focused on contactless security chips (e.g. for passports) and on the other has a strong focus on automotive applications for sensors (drive train, safety, e-mobility, tire pressure monitoring system). The development effort together with the highly professional and experienced team has led to a top position amongst the international competitors. In 2007 the VDA entrusted Infineon and Bosch GmbH to develop a VDA controller and set the standard in the market.

Infineon Technologies AG (IFAG) www.infineon.com	

Infineon is a world leader in semiconductors. Combining entrepreneurial success with responsible action, at Infineon we make the world easier, safer and greener. Barely visible, semiconductors have become an indispensable part of our daily lives. Chips from Infineon play an essential role wherever energy is generated, transmitted and used efficiently. Furthermore, they safeguard data communication, improve safety on roads and reduce vehicles' CO_2 emissions.

Infineon designs, develops, manufactures and markets a broad range of semiconductors and systems solutions. The focus of its activities is on automotive electronics, industrial electronics, communication and information technologies and hardware-based security. The product range comprises standard components, customer-specific solutions for devices and systems, as well as specific components for digital, analogue, and mixed-signal applications. About 60% of Infineon's revenue is generated by power semiconductors, about 20% by embedded control products (microcontrollers for automotive, industrial as well as security applications), and the remainder by radio-frequency components, sensors and other product categories. Infineon generates 33% of its revenue in Europe, 54% in Asia, and 13% in the Americas.

Infineon was founded in 1999, reported about 36,300 employees world-wide and €6,473 million revenue in the 2016 fiscal year (as of September 30, 2016) and is present with 34 research and development locations, 19 manufacturing locations, and about 45 sales offices worldwide.

The **Automotive segment (ATV)** designs, develops, manufactures and markets semiconductors for use in automotive applications. ATV offers a very broad product portfolio of microcontrollers, magnetic and pressure sensors, radio-frequency and particularly radar components as well as power semiconductors (discretes, modules and ICs). Infineon is the leading provider of system solutions for automotive electronics, with the industry's most comprehensive portfolio of power semiconductors, sensors and microcontrollers. Following the guiding principle of "clean, safe and smart" ATV addresses the industry's current megatrends: Electro-mobility automated driving as well as connectivity and advanced security. ATV generated revenue of €2,651 million in the 2016 fiscal year. According to Strategy Analytics, Infineon was ranked number 2 in the 2015 automotive semiconductor market with a market share based on revenues of 10.4%.

IMOTIVE	

ICPT is a specialized engineering company focused on design and development of Battery Packs, electronics and drivetrain technologies for EVs. We provide solutions for Electric Motorsport, Hybrid Vehicles, EVs, Mining Industry and Aeronautics.

ICPT's goal is to introduce ecological and economical solutions for future generations. The company is focused on creation and implementation of new, innovative technologies, which will allow minimizing environment pollution and need for traditional fuels. Those technologies are mainly focused on:

- Advanced Battery Packs
- Battery management systems – embedded software and robust hardware
- Fuel cells and energy management systems
- Systems integration

Around 30 highly-skilled engineers employed in the R&D, Integration, Software, and Mechanics departments develop new applications and solutions for efficient battery storage and management systems used in the automotive industry.

As the market grows, ICPT provides constant maintenance of existing projects and develops both customized and standard solutions to cope with the new needs.

| MANZ |  |

Manz Italy has been established in April 2014 following the acquisition of the mechanical engineering division of KEMET (formerly Arcotronics).

Arcotronics was founded in 1962 in Sasso Marconi near Bologna and has set worldwide standards for over 40 years in production technologies for capacitors and for nearly 30 years for Lithium batteries. In 1986, Arcotronics pioneered into the battery market supplying winding machines for lithium primary cells. Thus, Manz Italy can draw on long-term experience in engineering, development, and fabrication of semiautomatic and automatic production equipment for Lithium-ion batteries.

85 highly qualified employees are working for Manz Italy in a total area of 5,100 square meters.

CWM

Chemnitzer Werkstoffmechanik GmbH (CWM) was founded in 1990. The company is a provider of research and development services in the field of materials mechanics. It has more than 20 years of research experience in the field of material characterization and reliability assessment of smart system assemblies, by stress tests and various methods of microscopy and physical failure analysis.

Mainly product and process developments of the CWM emerged in the field of digital image correlation for the analysis of shifts and deformations. These are a key technology for CWM. CWM develops and distributes software systems (VEDDAC) and measuring systems (microDAC®) for digital image correlation, that are used by more than 50 national and international customers with a focus on micro technologies and materials science.

About the Editors

Eric Armengaud received his M.Sc. from ESIEE Paris, in 2002, the Ph.D. degree from the TU Vienna, in 2008 and the MBA degree from IBSA, in 2016. He has more than 15 years of experience in automotive embedded systems in different positions. He is currently project manager R&D with the responsibility to identify, set-up and manage national and European R&D programs such as INCOBAT. Eric Armengaud is author and co-author of more than 70 peer reviewed publications and patents, and is guest lecturer at the University of Applied Sciences FH Joanneum.

Riccardo Groppo took his M.Sc. degree in Electronic Engineering at the Politecnico of Torino (Torino, Italy). He started his career at Honeywell (1987), as a HW designer for workstation and mini-computer. He worked at Centro Ricerche FIAT (CRF) for almost 25 years, starting as a HW designer (1989) and then being the Head of the Automotive Electronics Design and Development Dept. (2002–2013). He has been directly involved in the industrial development of several breakthrough innovations in the powertrain (e.g. Common Rail, MultiAir, Dry Dual Clutch) in cooperation with the most relevant automotive suppliers and semiconductor manufacturers. He is the co-founder and CEO of Ideas & Motion (2013) and holds 29 patents in the field of automotive electronics and smart systems, most of which are currently in mass production on passenger cars. He is the Chairman of the "Transportation" Working Group within EPoSS and member of the European Advisory Council of the SAE International.

Sven Rzepka is Head of the Micro Materials Center (MMC) at Fraunhofer ENAS and Professor for 'Smart Systems Reliability' at TU Chemnitz, Germany. He joined Fraunhofer in 2009 after working as Principal Simulation at Qimonda, Backend development, and at Infineon Technologies, BEoL reliability department. In 2002, he graduated from TU Dresden with Diploma, Ph.D., and habilitation degrees. In total, Dr. Rzepka has been working in BEoL and packaging technologies for 30 years with 25 years of experience in microelectronics and smart systems reliability tests, analysis, and simulation. He has published his engineering work in more than 150 papers and technical talks in international journals and at conferences around the world, respectively, and received more than 10 best and outstanding paper awards. He is involved in international conference committees, e.g., of ASME InterPACK, EPoSS Smart Systems Integration, EuroSimE, and organizes the annual European Expert Workshop on Reliability of Electronics and Smart Systems (EuWoRel). Prof. Rzepka is member of IEEE for 22 years, and Euceman for 9 years, and ASME. Serving as member of the Executive Committee of the European Technology Platform EPoSS, he has frequently be involved in evaluation, road mapping, and policy making processes advising European Commission and national authorities.